WEST BEND LIBRARY

the TECHNOLOGY beHIND
SPORTS AND SPORTING EQUIPMENT

Nicolas Brasch

- How Has Technology Changed Running Shoes?
- Why Do Golf Balls Have Dimples?
- How Do High-Tech Swimsuits Give Competitors an Unfair Advantage?

A+

WEST BEND LIBRARY

This edition first published in 2011 in the United States of America by Smart Apple Media. All rights reserved. No part of this book may be reproduced in any form or by any means without written permission from the publisher.

Smart Apple Media
P.O. Box 3263
Mankato, MN, 56002

First published in 2010 by
MACMILLAN EDUCATION AUSTRALIA PTY LTD
15–19 Claremont St, South Yarra, Australia 3141

Visit our web site at www.macmillan.com.au or go directly to www.macmillanlibrary.com.au

Associated companies and representatives throughout the world.

Copyright © Nicolas Brasch

Library of Congress Cataloging-in-Publication Data

Brasch, Nicolas.
 Sports and sporting equipment / Nicolas Brasch.
    p. cm. — (The technology behind)
 Includes index.
 ISBN 978-1-59920-570-0 (library bound)
 1. Sports—Technological innovations—Juvenile literature. 2. Sporting goods—Juvenile literature. I. Title.
 GV745.B73 2011
 688.7'6—dc22
                                                                                                2009054436

Publisher: Carmel Heron
Managing Editor: Vanessa Lanaway
Editor: Georgina Garner
Proofreader: Erin Richards
Designer: Stella Vassiliou
Page layout: Stella Vassiliou and Raul Diche
Photo researcher: Wendy Duncan (management: Debbie Gallagher)
Illustrators: Guy Holt, p. 10; Alan Laver, pp. 7, 12, 13, 19, 20, 21; Richard Morden, pp. 8, 14, 29, 30;
   Karen Young, p. 1 and Try This! logo.
Production Controller: Vanessa Johnson

Manufactured in China by Macmillan Production (Asia) Ltd.
Kwun Tong, Kowloon, Hong Kong
Supplier Code: CP March 2010

Acknowledgements
The author and the publisher are grateful to the following for permission to reproduce copyright material:

Front cover photographs:
Bikes © Shutterstock/Arthur Eugene Preston; Shoes © Tobias Titz/Getty Images; Computer simulation © Colin Anderson/Blend Images/Corbis.

© Colin Anderson/Blend Images/Corbis, **9** (bottom); © Bettmann/Corbis, **26**; © Schlegelmilch/Corbis, **16** (bottom); Flickr/geishaboy500, **6** (bottom left); Flickr/davesneakers, **6** (bottom right); © Nicolas Asfouri/AFP/Getty Images, **8**; © Ian Barrett/AFP/Getty Images, **17** (top); © Paul Gilham/Getty Images, **27**; © Gavin Lawrence/Getty Images, **16** (top left); © Jamie McDonald/Getty Images, **14**; © Thinkstock/Getty Images, **11**; © Tobias Titz/Getty Images, **6** (top right); © Toru Yamanaka/AFP/Getty Images, **30**; courtesy of Henselite (Australia) Pty Ltd, **15** (all); © Chris Schmidt/iStockphoto, **4**; © walik/iStockphoto, **22**; MEA Images/Brand X, **23** (right); Newspix/Gregg Porteous, **9** (top); photolibrary/Dietmar Plewka, **6** (top left); photolibrary/GetMappingPLC, **17** (bottom); © Shutterstock/Sergei Bachlakov, **18** (right); © Shutterstock/Racheal Grazias, **28**; © Shutterstock/Anthony Hall, **25**; © Shutterstock/Junker, **21**; © Shutterstock/Konstantin Komarov, **23** (left); © Shutterstock/Arthur Eugene Preston, **5**; © Shutterstock/Eduard Stelmakh, **18** (top left); © Shutterstock/Peter Weber, **24**.

While every care has been taken to trace and acknowledge copyright, the publisher tenders their apologies for any accidental infringement where copyright has proved untraceable. Where the attempt has been unsuccessful, the publisher welcomes information that would redress the situation.

The publisher would like to thank Heidi Ruhnau, Head of Science at Oxley College, for her assistance in reviewing manuscripts.

Please note
At the time of printing, the Internet addresses appearing in this book were correct. Owing to the dynamic nature of the Internet, however, we cannot guarantee that all these addresses will remain correct.

Contents

What Is Technology?	4
The Technology Behind Sports and Sporting Equipment	5
How Has Technology Changed Running Shoes?	6
Why Do Today's Athletes Run Faster Compared to 100 Years Ago?	8
How Does a Heart-Rate Monitor Work?	10
How Does Hawk-Eye Judge Close Line-Calls in Tennis Matches?	12
What Makes Lawn Bowls Travel in a Curve?	14
Has Technology Made Auto Racing Safer?	16
Why Do Racing Cars Have Smooth Tires?	18
Why Do Golf Balls Have Dimples?	20
How Does a Snowboard Work?	22
What Makes Racing Bikes Faster than Normal Bikes?	24
How Do Performance-Enhancing Drugs Work?	26
How Do Pole-Vaulters Launch Themselves So High?	28
How Do High-Tech Swimsuits Give Competitors an Unfair Advantage?	30
Index	32

Look out for these features throughout the book:

"Word Watch" explains the meanings of words shown in **bold**

"Web Watch" provides web site suggestions for further research

What Is Technology?

> We're changing the world with technology.
>
> Bill Gates, founder of the Microsoft Corporation

The First Tools
One of the first examples of technology, where humans used their knowledge of the world to their advantage, was when humans began shaping and carving stone and metals into tools such as axes and chisels.

science knowledge that humans have gathered about the physical and natural world and how it works

▲ People use technology every day, such as when they turn on computers. Technology is science put into action to help humans and solve problems.

Technology is the use of **science** for practical purposes, such as building bridges, inventing machines, and improving materials. Humans have been using technology since they built the first shelters and lit the first fires.

Technology in People's Lives

Technology is behind many things in people's everyday lives, from lightbulbs to can openers. It has shaped the sports shoes people wear and helped them run faster. Cars, trains, airplanes, and space shuttles are all products of technology. Engineers use technology to design and construct materials and structures such as bridges, roads, and buildings. Technology can be seen in amazing built structures all around humans.

Technology is responsible for how people communicate with each other. Information technology uses scientific knowledge to determine ways to spread information widely and quickly. Recently, this has involved the creation of the Internet, and e-mail and file-sharing technologies. In the future, technology may become even more a part of people's lives, with the development of robots and artificial intelligence for use in business, in the home, and in science.

The Technology Behind Sports and Sporting Equipment

The drive to be number one in sports has driven athletes, coaches, companies, and even nations to spend huge amounts of money trying to get an edge. Much of this money is spent on technology to improve equipment, clothing, and training techniques.

Fair or Unfair?

Technology is a driving force behind sports and games. It put the dimples in golf balls, and it created the snowboard.

Many technological **innovations** have been good for competitors, spectators, coaches, and officials, as well as for sports and games in general. Auto racing, for example, is a sport that has used technology to improve safety for competitors. The introduction of Hawk-Eye, an electronic signaling **device**, has made tennis fairer.

Some technology, however, is controversial. Sporting officials must now deal with illegal performance-enhancing drugs that give their users unfair advantage and terrible long-term side effects. **High-tech** swimsuits are another technology that is controversial, although not harmful.

▼ Technology has helped cyclists, runners, and swimmers move faster and break more records. These cyclists' helmets, eyewear, clothing, and bikes use technological innovations.

The People Behind the Technology

Many people with different jobs are behind the technology of sports and sporting equipment.

Sports Physiologist Understanding how the human body performs and working out ways to better an athlete's performance

Sports Nutritionist Understanding the role nutrition plays in performance

Computer Scientist Designing the software for technologies such as Hawk-Eye

Sports Architect Designing sporting arenas and circuits such as auto-racing tracks

Word Watch

device something made for a particular purpose or to do a particular job

high-tech (high technology) advanced

innovations new ideas and new ways to do things

How Has Technology Changed Running Shoes?

There was a time when Olympic athletes simply chose running shoes that were comfortable. Some long-distance athletes, particularly from African nations, even ran in bare feet. Today's top-of-the-range running shoes are designed to adapt to each athlete's individual needs.

History of Running Shoe Technology

The first running shoes, called plimsolls, were developed in Britain in the 1800s. For many years, they were only available in black or white.

A breakthrough in running shoes occurred in the 1890s when British company W. Foster and Sons (now known as Reebok) developed running shoes with spikes in the soles. A German man, Adi Dassler, began to produce different running shoes with different types of spikes for different distances. His company became known as Adidas.

During the 1970s, the company Nike had two technological breakthroughs. They developed running shoes with a "waffle" sole and also developed running shoes that had a gas-filled bag of air inserted in the sole. Some of the Nike air bubble shoes, particularly those used for basketball, came with a special pump valve so that more air could be added when needed.

Plimsoll Sneakers

Plimsolls were named after the Plimsoll line, which is the name given to a ship's waterline, because the line around the top of the sole of the plimsolls reminded people of this line. Plimsolls became known as sneakers because people could sneak around in them without being heard.

◀ Plimsolls had a canvas upper with a rubber sole. They were comfortable for running and doing exercise.

▶ Adding spikes to the soles of shoes gives athletes more **traction**.

◀ A waffle pattern on rubber soles provides good grip. The Nike designer who came up with this idea actually burned some rubber in a waffle iron to test the idea.

▶ Air soles help cushion an athlete's feet from the impact of running. The air-cushioning unit can be seen through the side of the sole.

Word Watch

traction grip

Inside a High-Tech Running Shoe

The most **high-tech** running shoes today use a **microprocessor**, sensors, and a **mechanical** cable system that work together to adjust the cushioning of the shoe. The microprocessor takes into account the runner's size, pace, **fatigue** level, and the type of ground the person is running on, and adjusts the cushioning to reduce the chances of a strain or injury. High-tech running shoes use technology to adjust again and again to a runner's needs.

- control buttons
- microprocessor
- sensor system
- cable system
- motor

The microprocessor is programmed to respond to different readings. It is capable of doing 5 million calculations per second.

The microprocessor triggers the mechanical cable system to adjust the cushion in the heel to match the conditions and the runner's needs.

As the runner's foot hits the ground, the sensor system in the heel measures the **compression** and sends readings to the microprocessor. The sensor can take up to 1,000 readings per second.

High-Tech Soccer Shoes

Today's top soccer players can have shoes made especially for them with a certain number of layers of material that takes into account their blood flow. This makes the shoes suitable for both cold and warm conditions.

Millions of Sports Shoes

More than 350 million pairs of sports shoes are sold in the United States each year. This is more than one pair a year for every man, woman, and child in the country.

Word Watch

compression pressing things together into a smaller space

fatigue extreme tiredness

high-tech (high technology) advanced

mechanical related to machines, machinery, and physical work

microprocessor very small controller that has an electronic circuit

Why Do Today's Athletes Run Faster Compared to 100 Years Ago?

If you look at past world records for 300-foot (100-m) races, you can see that today's athletes are running much faster. This difference is a result of advances in technology in areas such as training, nutrition, and equipment.

Comparing Now and Then

At the 2008 Olympic Games, Usain Bolt won the 100-meter running race in a world record time of 9.69 seconds. At the first modern Olympic Games, 112 years earlier, Thomas Burke won the 100-meter final in 12.0 seconds. If Thomas Burke had been in the same race as Usain Bolt, he would only have been at the 80-meter mark as Bolt crossed the finish line.

Comparing Marathon Times

In 1896, the marathon, a distance of 26.219 miles (42.195 km), was won in a time of 2 hours, 58 minutes, 50 seconds. In 2008, the marathon was won in 2 hours, 6 minutes, 32 seconds. If they were in the same race, the winner of the 1896 marathon would have been around the 19-mile (30-km) mark at the moment the 2008 winner was crossing the finish line.

◀ Jamaican athlete Usain Bolt broke the world record when he won the men's 100-meter final at the Beijing Olympic Games in 2008.

▼ If all the winners of the men's 100-meter final at the Olympic Games were in the same race, this is where they would finish.

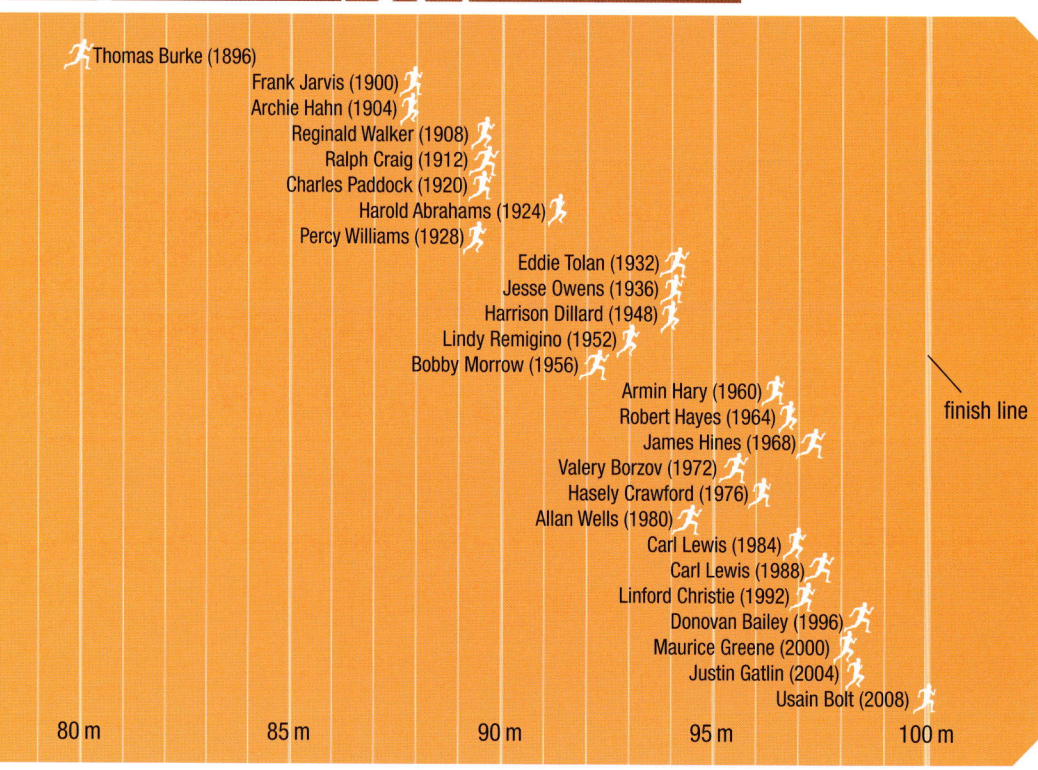

Web Watch
www.olympic.org/en/content/Sports/All-Sports/Athletics/Track/All-Track-events/100mMen

Video Technology

One-hundred years ago, an athlete's coach or trainer would watch from the side of the running track and then advise the athlete how to alter his or her movements to gain more speed. Today, footage of an athlete can be paused and slowed down so that the trainer and athlete can **analyze** every part of the athlete's body and every moment in a race.

◀ Athletes can be videotaped so they can watch themselves running and spot anything they could do better. They can then make adjustments to their movement to gain more speed.

Computer Technology

Computer **simulation** and modeling involve creating a model of something on a computer and then making it act and react in certain ways. In sports, computer simulation and modeling can be used to determine how an athlete can improve.

▶ Computer simulations give athletes tips on how to move their bodies more quickly and efficiently.

Nutrition and Technology

Food technology looks at the makeup of different foods and how they affect the body. Sprinters require strong muscles to help them power out of the blocks and to **propel** them over short distances. The best foods for building muscles are those rich in **protein**, such as red meat, eggs, and milk. Long-distance runners require food that will give them **stamina**, without building up their muscles too much. The best foods for stamina are those rich in **carbohydrates** and fat, such as pasta, cereals, fruit, vegetables, and oil.

Quicker Starts

Another reason sprinters are much faster today is the way races are started. At the first modern Olympic Games, most of the 100-meter runners stood upright at the start, although a couple crouched down. Today, all sprinters start from a low position, to give them more propulsion. Today's sprinters also push themselves off the starting blocks.

Word Watch

analyze examine something in detail
carbohydrates chemical compounds produced by plants that serve as an energy source for animals
propel push forward
protein chemical compounds that are made up of amino acids and are an essential part of all living things
simulation computer model or something else that pretends to be real
stamina staying power

How Does a Heart-Rate Monitor Work?

A heart-rate monitor measures a person's heart rate, which is the amount of times the person's heart beats in a minute. Some athletes use heart-rate monitors when they train to check that their hearts are working hard, circulating blood around their bodies.

How the Heart Works

The heart is a muscle that pumps blood around the body by squeezing and relaxing. The right atrium receives blood from the body and passes it into the right ventricle, which pumps the blood into the lungs. The left atrium receives blood from the lungs and passes it into the left ventricle, which then pumps the blood into the body.

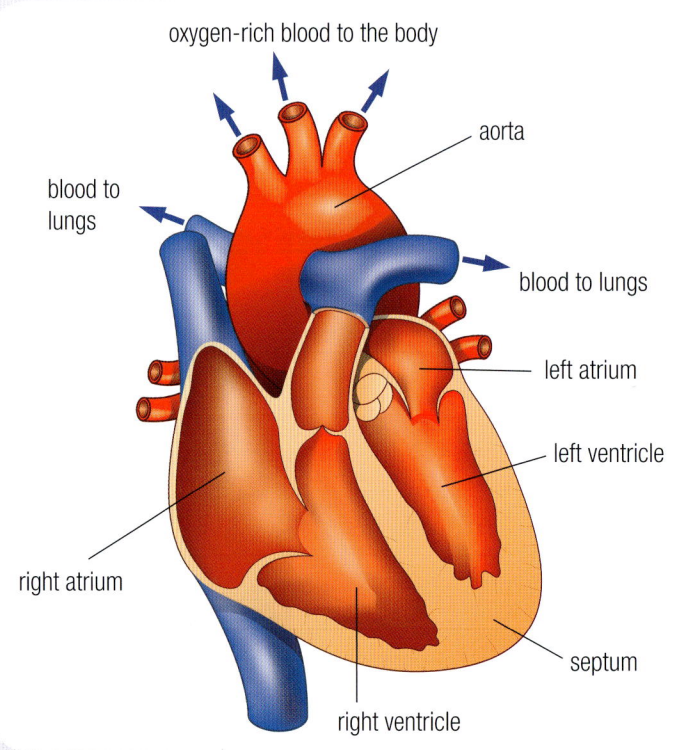

◀ The heart has four chambers. The left and right ventricles are at the bottom of the heart, and the left and right atriums are at the top of the heart. A thick wall of muscle, called the septum, separates the right-hand side from the left-hand side.

History of the Heart-Rate Monitor

Heart-rate monitors existed before 1977, but they were large and awkward and could not be used without disrupting an athlete's normal training pattern. In 1977, Finnish company Polar Electro invented the first wireless heart-rate monitor. It was created for the Finnish cross-country ski team, so that the team could monitor its training program and make necessary changes. By the 1990s, improvements in technology had made heart-rate monitors affordable for amateur runners who just wanted to monitor their own progress.

BPM

A person's heart rate is measured in bpm, which stands for "beats per minute." There are different healthy heart rate ranges or target ranges for different ages.

Try This!

Measure Your Own Heart Rate

You can compare your heart rate before and after exercise by measuring the beat of your pulse.

1. Before you start exercising, find your pulse at your wrist. Place your index and middle fingers on your pulse and count how many pulse beats occur in 10 seconds.

2. Multiply this number by six to get your heart rate per minute.

3. Run 330 feet (100 m).

4. Repeat steps 1 and 2, and compare the two heart rates.

Web Watch

kidshealth.org/kid/htbw/heart.html

www.pbs.org/americaswalking/gear/gearheart.html

Transmitting and Receiving Heart-Rate Information

The main parts of a heart-rate monitor are the **transmitter** and the **receiver**. The transmitter has **electrodes** that record the electrical activity of the heart. These electrodes need to be positioned just below the breast. Each time the person's heart beats, the electrodes detect the movement and convert the movement into a radio signal. This signal is sent to the receiver.

The receiver receives the signals, measures the intervals between them over a short period of time, and then calculates the average interval. It calculates the user's heart-rate based on the average, and then it displays the results for the person to see.

Other Features

Basic heart-rate monitors only measure and display a person's heart rate, but more advanced models display other measurements and have features such as:

- an alarm that alerts users when they have reached their desired heart rate
- a timer that indicates how much longer the user has to exercise
- display that shows the number of calories burned during an exercise session
- the capability to download a personalized training program from a computer
- the capability to upload personal information to a computer
- a light so that the results can be easily read during early morning and night exercise sessions

▲ The transmitter is attached to a strap that is wound and tied around the chest, directly against the skin. The receiver is usually a watch-like **device** that is worn around the wrist.

Things that Affect Heart Rate

It is not just the rate of exercise that determines someone's heart rate. Other factors are:
- heat (the heart rate is usually faster in hotter conditions)
- **dehydration** (the heart works harder and beats faster when people are dehydrated)
- time of day (heart rate is usually lower earlier in the day)

Word Watch

dehydration lack of water

device something made for a particular purpose or to do a particular job

electrodes parts of an electrical system that collect electrical signals

receiver something that receives a signal

transmitter something that sends a signal

How Does Hawk-Eye Judge Close Line-Calls in Tennis Matches?

Hawk-Eye is the name given to the technology that detects whether or not a tennis ball lands in or out of the court. This technology is also used in some other sports, such as cricket.

Watching Like a Hawk

Hawk-Eye was invented in 2001 by British engineer Paul Hawkins. The name, Hawk-Eye, was partly inspired by his surname and also because a hawk has very good eyesight. Hawkins came up with the idea for Hawk-Eye to help cricket umpires make correct decisions. He then developed the technology to make it suitable for tennis.

How Hawk-Eye Tracks the Path of the Ball

If a player disagrees with a decision from a line umpire, he or she can get an instant replay from Hawk-Eye.

▶ Hawk-Eye follows the ball using a number of high-speed video cameras placed high above the tennis court. The cameras track and measure the flight of every ball that is hit.

1 » The center of the ball is identified within the **frames** taken by each camera. The position of the lines on the court are considered to take into account the different positions and the movement of the cameras.

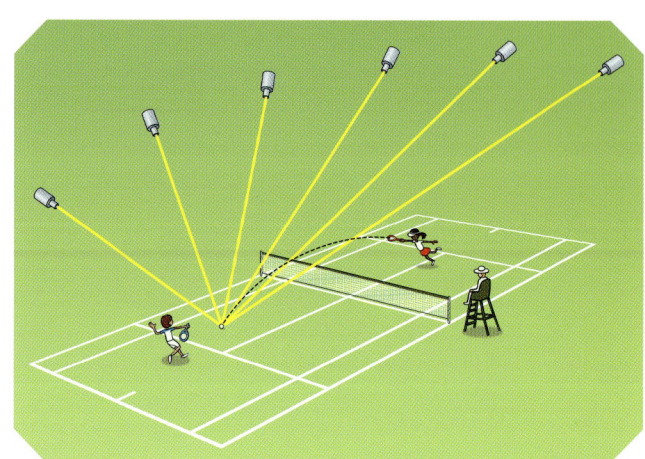

camera view A

camera view B

2 « The system **triangulates** the information from each camera to find the **three-dimensional** position of the ball.

3 » Steps 1 and 2 are repeated for each frame taken by the cameras. The three-dimensional positions of the ball are combined to find the **trajectory** of the ball.

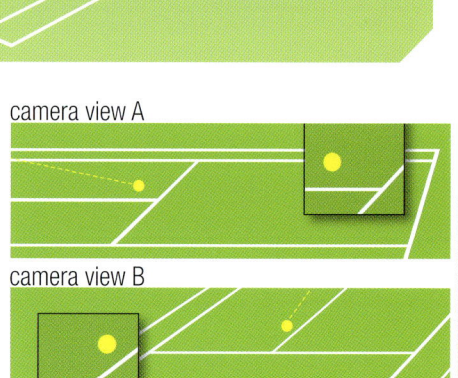

trajectory

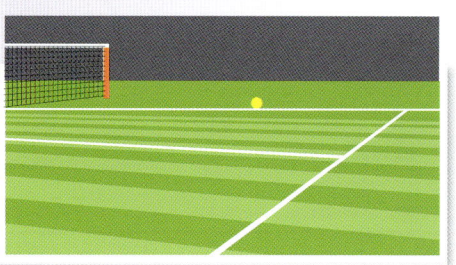

point of contact

4 « The trajectory is used to calculate exactly where the ball bounced and made contact with the court.

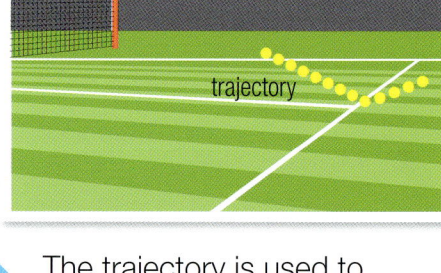

Word Watch

frames single pictures from a series of film footage
three-dimensional using length, width, and depth
trajectory curved path of an object in flight
triangulates determines a location using information and triangular shapes

12

Tennis Technology Before Hawk-Eye

Before Hawk-Eye was invented, tennis authorities tried two other **devices** to help improve the game of tennis and, in particular, the experience of those watching a game live.

Cyclops System

The Cyclops system was set up to help determine whether or not a serve was "in" or "out." It did not cover the outside lines of the court. It was named after a one-eyed monster in Greek mythology.

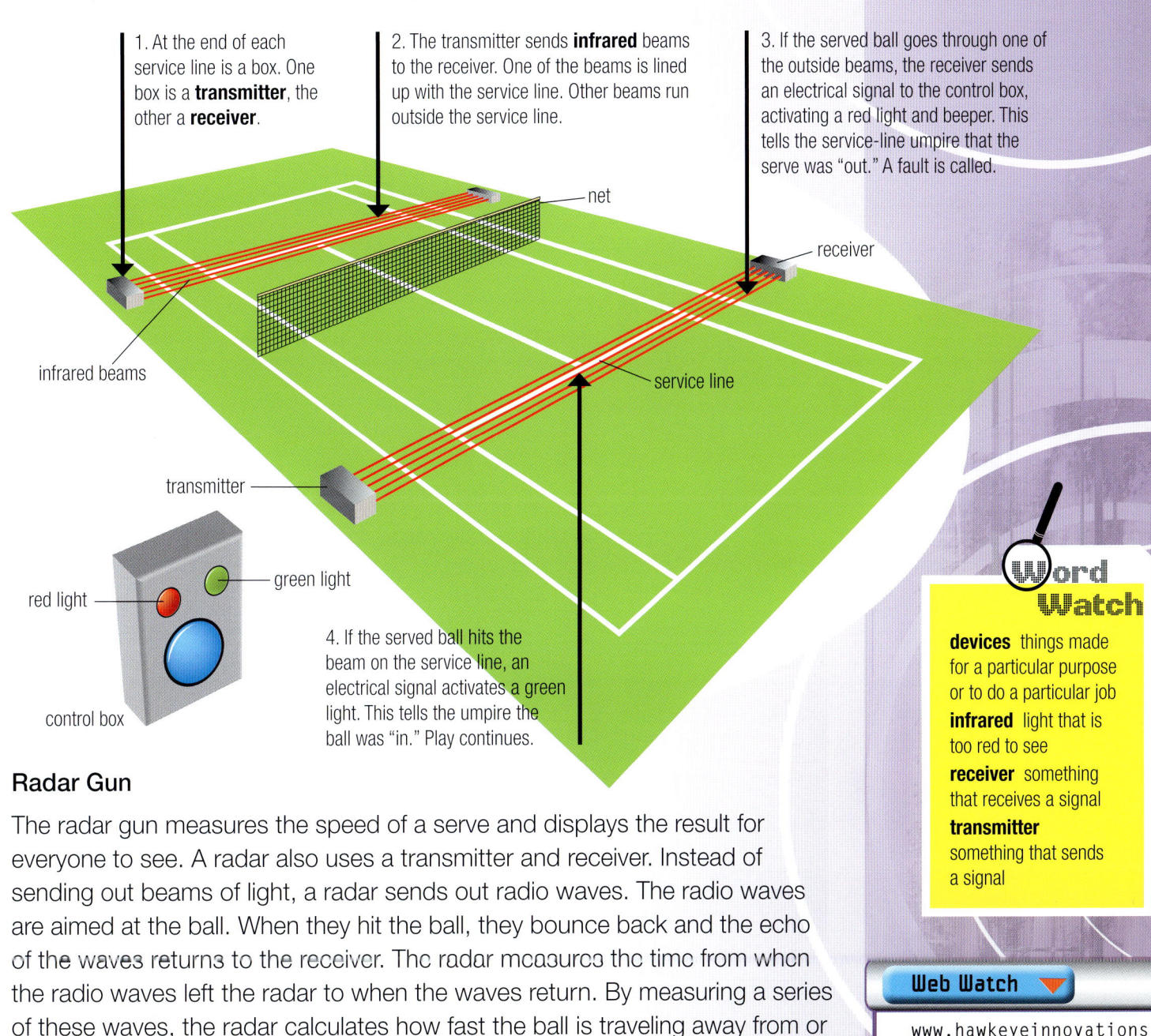

1. At the end of each service line is a box. One box is a **transmitter**, the other a **receiver**.

2. The transmitter sends **infrared** beams to the receiver. One of the beams is lined up with the service line. Other beams run outside the service line.

3. If the served ball goes through one of the outside beams, the receiver sends an electrical signal to the control box, activating a red light and beeper. This tells the service-line umpire that the serve was "out." A fault is called.

4. If the served ball hits the beam on the service line, an electrical signal activates a green light. This tells the umpire the ball was "in." Play continues.

Radar Gun

The radar gun measures the speed of a serve and displays the result for everyone to see. A radar also uses a transmitter and receiver. Instead of sending out beams of light, a radar sends out radio waves. The radio waves are aimed at the ball. When they hit the ball, they bounce back and the echo of the waves returns to the receiver. The radar measures the time from when the radio waves left the radar to when the waves return. By measuring a series of these waves, the radar calculates how fast the ball is traveling away from or toward the radar. This is the speed of the serve.

Word Watch

devices things made for a particular purpose or to do a particular job
infrared light that is too red to see
receiver something that receives a signal
transmitter something that sends a signal

Web Watch

www.hawkeyeinnovations.co.uk

13

What Makes Lawn Bowls Travel in a Curve?

Lawn bowling is a sport that is usually played on a closely mown lawn of grass, called a green. It is played with heavy balls, called bowls. When rolled along the green, a bowl does not roll in a straight line. It rolls in a curve. This is because of the way it is made.

The Bias of the Bowl

Lawn bowls are rounded on one side and oval shaped on the other. The **center of gravity** is not in the middle of the bowl, but slightly to the side. This means that the bowl rolls toward the center of gravity when it is moving. This movement is called bias.

Knocking Out the Competitors

A bowl that is rolled very hard may not turn at all. Some bowlers who wish to knock other bowls out of the way aim directly at these bowls and roll their bowls as fast as they can.

▲ When viewed from the front, a bowl appears oval.

Playing the Game

The aim of lawn bowling is to move your bowls as close as possible to a smaller ball, called the jack or kitty. Players take turns rolling their bowls.

A player needs to know which way the bowl is going to roll before making a shot. Manufacturers of lawn bowls put their emblem or another symbol on the sides of the bowls, along with information about the size of the ball. One of these emblems is smaller than the other and this signifies the side that the bowl will turn toward when it is rolling. A bowler who accidentally holds and rolls the bowl the wrong way will see it turn away from the other bowls and roll completely off the green.

▶ A bowler has to consider the bias of the bowl and how far he or she is bowling. The longer a bowl is rolling, the sharper the turn it will take.

center of gravity point where an object's or person's weight will evenly balance

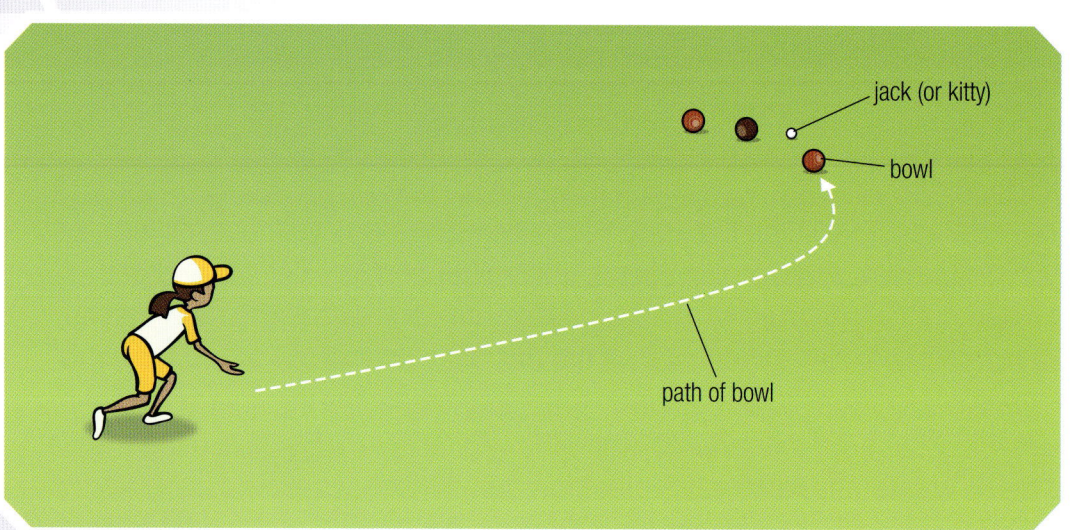

14

Making Lawn Bowls

The bias is created when the lawn bowls are made. The process of making lawn bowls involves several steps.

1 » Special powders are ground and mixed together. The powder mixture is then packed into a mold.

2 » The mold is placed in a machine that both presses and heats the mold. The object that comes out is in the shape of a bowl. It is known as a blank.

3 » The blank is left to cool, before it is inspected and weighed. Specially designed equipment turns, smooths, and polishes the blanks to make sure they are completely smooth and are exactly the right weight and size.

4 » The sides of the bowls are engraved.

5 » The bowls are then painted.

6 » The bowls are tested on a special table to make sure they have exactly the right bias.

An Ancient Game
Games similar to lawn bowling have been around for thousands of years. The ancient Egyptians, Greeks, and Romans all had their own versions.

Modern Lawn Bowling
Modern lawn bowling is associated with the British version, which was played as long ago as the 1200s. It is said that in 1588, the British sea captain Sir Francis Drake was lawn bowling as the Spanish fleet approached Britain. He insisted on finishing his game before sailing out to fight the enemy.

Has Technology Made Auto Racing Safer?

Auto racing is a dangerous sport, because it involves driving cars that are full of fuel at high speeds. However, changes are constantly being made to the cars, the tracks, and the rules to make the sport as safe as possible.

Reducing the Risks

Before the 1960s, it was generally accepted that deaths would occur in auto racing. But in the 1960s, racing authorities started changing their attitude. They acknowledged that accidents would happen, but they decided to do everything possible to avoid deaths from these accidents.

Harnesses and Seat Belts

One of the first changes was to make sure that drivers wore harnesses or seat belts. Before 1968, very few drivers wore these protective **devices**. Most people believed that being thrown from a car was safer than being strapped inside with the car leaking fuel. Research showed that it was safer to wear a seat belt or harness.

Fire Extinguisher Systems

Fire extinguisher systems were installed in racing cars. The problem with these early systems was that they required drivers to turn them on. Automatic systems with sensors were developed. The sensors detect fire and automatically activate the extinguishers.

First Auto Race
The first official auto car race took place in France in 1894 between the capital, Paris, and the town of Rouen, over a distance of 78 miles (126 km).

◀ A harness restrains a racing car driver so he is not thrown forward in an accident.

▼ The combination of fuel and heat makes racing cars a very high fire risk.

Word Watch

devices things made for a particular purpose or to do a particular job

Protective Clothing

Drivers' clothing has also been improved so that it is as fire **resistant** as possible. Technological advances in this area have revolved around the materials used in the suits. The multilayered suits must also be comfortable for drivers, who are squashed into a small, hot compartment for several hours.

▶ Driving suits are made from materials that resist fire. A helmet also protects the driver.

Worst Accident
The worst accident in auto racing occurred at the Le Mans track in France in 1955. A car driven by Pierre Levegh hit the rear of another car, became airborne and landed among the crowd, splitting into several pieces. The fuel tank burst into flames. More than 80 people were killed, including the driver.

Changes to Racetracks

Technology has also been used to redesign racetracks. The spectators' viewing area was moved back from the track. This provided drivers with more space to try to control their cars if they ran off the track.

Another important change was the inclusion of more corners and **chicanes**. Track changes have been made easier since the introduction of computer **simulation** technology. Track designers can work out exactly what speeds are achievable on each stretch of road before they build the racetrack.

▶ More corners and chicanes means less long straights where drivers can reach speeds of more than 185 miles (300 km) per hour. Drivers are in great danger if they lose control at these fast speeds.

Studying Crashes

Whenever a major auto racing accident occurs, authorities try to determine how the accident occurred and what could have been done to prevent it. Crash recorders in the cars, similar to the black box used in airplanes, help determine what happened. The information collected is evidence of how the car behaved in the seconds before the crash.

Word Watch

chicanes short stretches of road that twist and turn
resistant able to withstand the effects of something
simulation computer model or something else that pretends to be real

17

Why Do Racing Cars Have Smooth Tires?

Normal cars use tires with treads. Treads are patterns that are cut into the tire and run around the whole tire. Racing cars usually use completely smooth tires, called slicks.

Tire Treads

Tread patterns on tires stop a car from sliding around on wet and slippery surfaces. They also soak up water and dispose of the water behind the car as the car moves along.

On a dry surface, there is no need for a tire with treads. If a car race is going to take place on a wet surface, the racing cars are fitted with treads. Most car races take place on a dry track and so the racing cars use slicks.

Racing Slicks

Racing cars need to grip the road tightly. Good grip makes it easier and safer for the driver to steer the car around corners without losing control at high speeds. The best possible grip requires as much of the surface of the tire as possible to be in contact with the surface of the track. This is why racing cars usually use smooth tires.

If it starts raining during a race, cars have to return to the **pit** and have their tires replaced. A car that continues racing with slicks on a wet track runs the risk of aquaplaning. Aquaplaning is when a car is lifted off the road by a buildup of water in front of and underneath the tire.

Drag Racing

The first type of auto racing to use slicks was drag racing. Drag racing involves cars racing over a short, straight course. Drag races last only a few seconds, so getting a good start is very important.

▲ Tread tires have cuts and grooves that stop them sliding in wet conditions.

Word Watch

pit area of a racetrack where cars are refueled and have their tires replaced

▲ Slicks do not have cuts and grooves, so they have a larger contact area with the road and therefore better grip. Pit crews must work quickly to fit slicks on a car during a race.

How Tires Are Made

A tire is made up of several layers and parts. The various layers are put into a machine and pressed together. Their ends are **fused** using extreme heat.

The tire is then put into another machine. It is put under great pressure from all angles, has the tread branded into it, and has the outside heated to toughen it. Finally, the tire is cooled and checked.

Parts of a Tire

A tire gains its strength from many different layers.

Filling the Tires

Regular car tires are filled with air, but most racing car tires are filled with nitrogen gas. Nitrogen is used because it does not expand and contract as much as air, so it gives drivers a smoother ride.

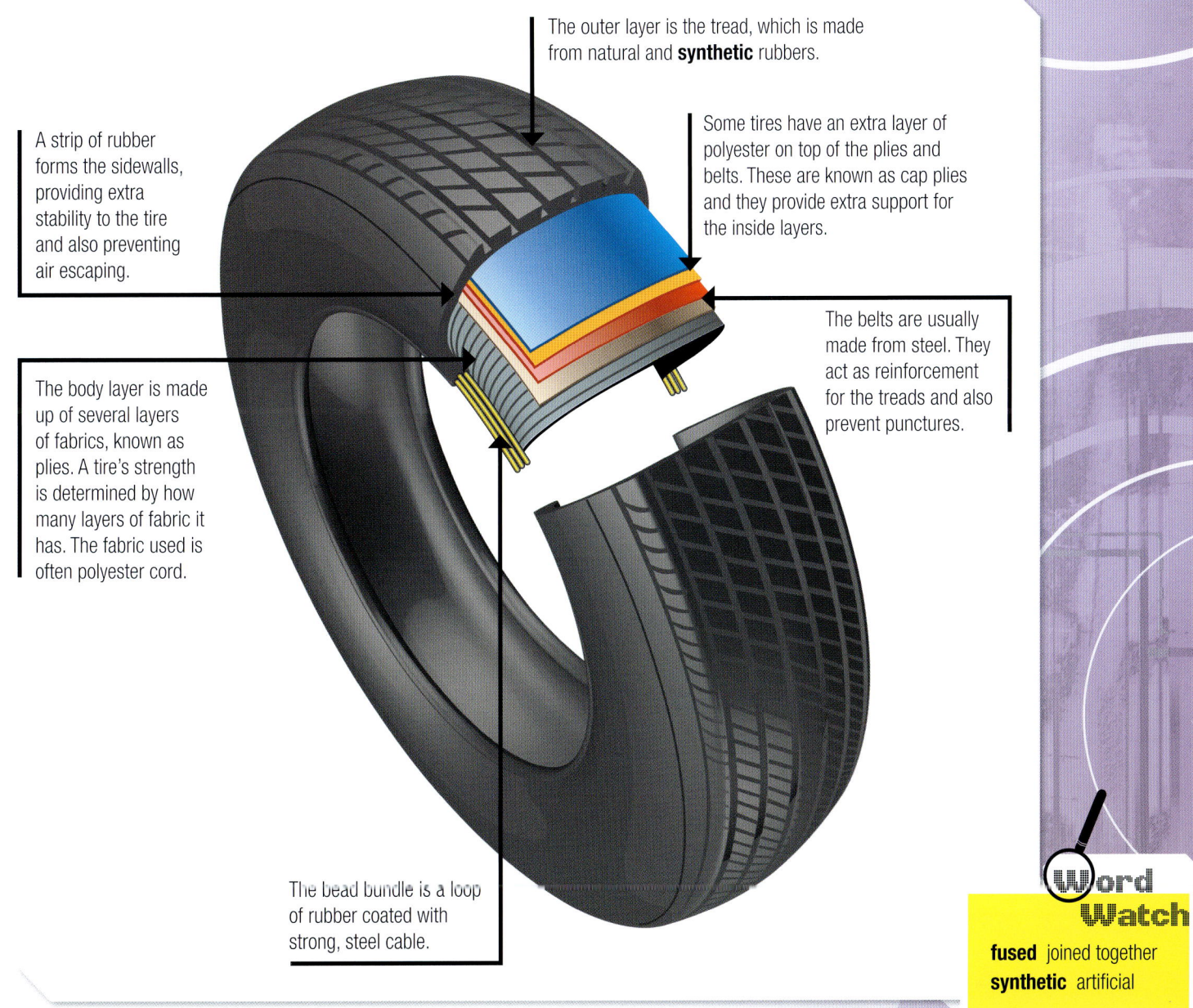

The outer layer is the tread, which is made from natural and **synthetic** rubbers.

Some tires have an extra layer of polyester on top of the plies and belts. These are known as cap plies and they provide extra support for the inside layers.

A strip of rubber forms the sidewalls, providing extra stability to the tire and also preventing air escaping.

The belts are usually made from steel. They act as reinforcement for the treads and also prevent punctures.

The body layer is made up of several layers of fabrics, known as plies. A tire's strength is determined by how many layers of fabric it has. The fabric used is often polyester cord.

The bead bundle is a loop of rubber coated with strong, steel cable.

Word Watch

fused joined together
synthetic artificial

Why Do Golf Balls Have Dimples?

There are between 250 and 450 dimples on a golf ball. These dimples help golf balls travel farther when hit. This has to do with the scientific concepts of **drag** and **lift**.

Drag and Airborne Balls

There are two types of drag at work on an airborne ball. The first is a type of **friction**. Air hits the front of the ball, slowing it down a little.

The other type of drag has a larger effect. This drag occurs because the air tries to cling to the ball, but it separates behind the moving ball after it has passed over and under it. The separation causes **eddies** that slow down the ball. This action is called laminar flow.

First Dimpled Golf Balls

Golfers used smooth balls until they discovered, by accident, that old golf balls with scuff marks and dents traveled farther than new, smooth balls. The first dimpled golf balls appeared in the early 1900s.

▶ As a ball passes through the air, the airflow passes over and around the ball, separating behind the ball and creating small eddies that draw energy from the movement of the ball. This laminar flow happens almost immediately.

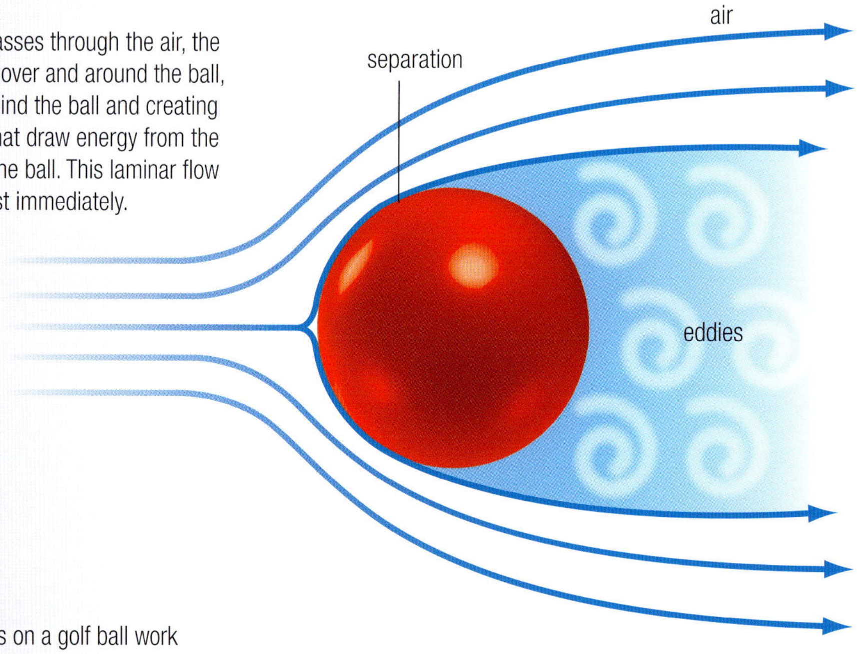

▶ The dimples on a golf ball work against laminar flow. The air clings to the dimples in the surface of the ball, the act of separation occurs later in the flight, and fewer eddies are produced. As a result, the ball travels farther. This is known as turbulent flow.

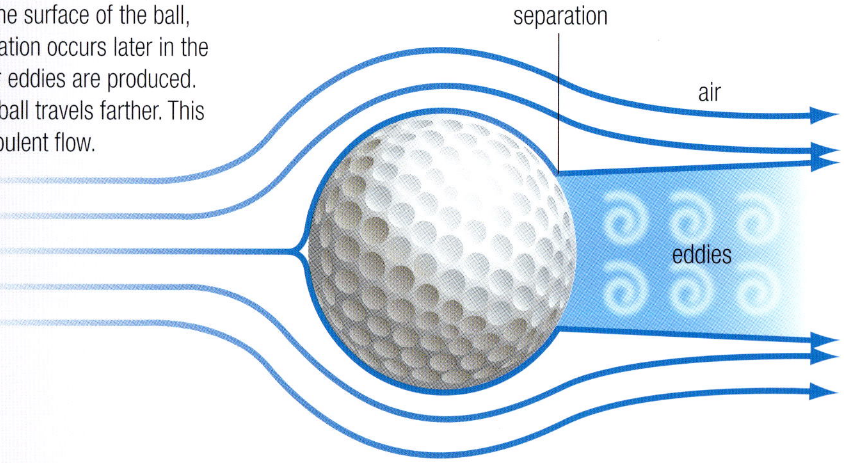

Word Watch

drag resisting force of air or liquid
eddies air currents with circular motion
friction force that resists motion and slows an object
lift an upward force

20

Lift and Airborne Balls

Dimples also improve the flight of golf balls by increasing their lift. Lift is an **aerodynamic** force that enables objects to fly through the air.

When a golf ball is hit, it spins backward as it moves through the air. It spins this way due to the angle of the golf club as the ball is hit. The backspin motion causes air to flow faster over the top of the ball than under the ball. Fast-flowing air has less pressure than slow-flowing air, so the pressure of the air underneath the ball helps keep the ball up for longer.

St Andrews, Home of Golf

Golf began in Scotland in the 1400s. Similar games, where objects were hit with clubs, were played in the Netherlands and Belgium around the same time, but it was the Scots who developed the idea of hitting a ball into a hole. The rules of golf were developed by the Royal and Ancient Golf Club of St. Andrews, Scotland.

▶ The dimples on a golf ball mean that the air clings to the ball longer than if it were a smooth ball, increasing the length of time that the ball experiences lift.

Golf Clubs and Technology

A golfer carries a range of different golf clubs. Some clubs are designed to hit the ball a long way, while others are used closer to the hole, where accuracy rather than distance is important.

Golf clubs have heads that are angled. The greater the angle of the head, the shorter the distance it is used for. The clubs that are used most of all are called irons. Irons are numbered. The higher the number, the greater the angle of its head.

▶ Golf clubs called drivers are used to hit the ball long distances. They have little angle because height is not as important as distance.

Word Watch

aerodynamic to do with the forces involved in flight

How Does a Snowboard Work?

Snowboarding is a sport that takes place on snow, but it is more similar to surfing and skateboarding than it is to snow sports such as skiing. Snowboards can be moved particular ways because of their shape and the materials they are made from.

Parts of a Snowboard

All snowboards are constructed with a few common parts, such as the core, base, and top sheet.

Moving the Board
One of the most important features of a snowboard is its camber. The camber is the curve between the front of the board (the nose) and the back of the board (the tail) when the board is lying flat. The springier the camber, the easier it is to move the board into different positions.

Bindings
Bindings on top of the snowboard keep the snowboarder's feet in place.

Top sheet
A plastic sheet covers the top layer of fiberglass and protects the inside of the board. The board's graphics and manufacturer's details are on this plastic sheet.

Core
The core is made from wood or foam. It gives the board its shape and strength. Steel edges are attached around the edges of the core to help the rider control the turn and speed of the board.

Layers
Fiberglass layers are **bonded** with **resin** to the top and bottom of the core. This provides extra stiffness and strength, while keeping the board as light as possible.

Base
The base is made from solid, smooth plastic so it can move quickly and easily on slippery surfaces. There are two types of bases. Extruded bases are basic, cheap models that are melted into shape. Sintered bases are more expensive and are made through a process of heating and pressure. They produce less **friction** between the board and the snow, which makes them faster than extruded models.

Word Watch
bonded stuck together
friction force that resists motion and slows an object
resin sticky substance similar to glue

22

Types of Boards

There are three types of snowboards:

- Freeride boards are best for beginners. They are easy to handle and work in all snow conditions.
- Freestyle boards are short and wide. They are designed for spinning and doing other tricks, but they cannot do quick turns.
- Freecarving boards are long and narrow. They are designed for speed down the slopes, as well as for quick turns, and they are used for downhill racing.

Tips for Riding a Snowboard

The first rule of snowboarding is to secure your feet into the bindings while on a level part of snow. You do not want to be sliding downhill while securing your feet to the board!

To stop the board, you should push down hard with both heels or toes, so that the board turns sharply and stops.

Popular Sport

Snowboarding began in the United States in the 1970s. It became popular so quickly that it was introduced as a sport at the Winter Olympic Games in 1998.

▼ Snowboarders control their boards by shifting their weight between their heels and toes and from foot to foot. This movement causes the edges of the board to dig into the snow, which causes the board to turn in particular directions.

▼ Snowboarders should keep their knees bent. This gives them more control over the board, particularly for turning. It also makes it easier to balance when going over bumps.

Web Watch

adventure.howstuffworks.com/snowboarding.htm
www.olympic.org/en/content/Sports/All-Sports/Skiing/Snowboard/Snowboard-Equipment-and-History

What Makes Racing Bikes Faster than Normal Bikes?

Riding a racing bike is very different from riding a normal bike. Racing bikes have many features that normal bikes do not—and they are also missing a lot of things that are considered essential on a normal bike.

Types of Racing Bikes

There are two types of racing bikes: track bikes and road bikes.

Track Bikes

Track bikes are used for racing around tracks with raised sides. These tracks are called velodromes. A track bike's frame needs to be strong and flexible because it undergoes enormous pressure during a race, with the rider twisting it around corners and past other riders. The bike must also be light, because a light bike will go faster than a heavy bike.

Track bikes have only one gear, which connects the rear wheel and the pedals. When the rear wheel is moving, the pedals and the cyclist's legs must move too. Gears add weight to bikes, so having just one gear makes the bike lighter than a multigeared bike.

Track bikes do not have brakes. They are designed for racing, not for stopping. Cyclists slow down after a race by pedaling slower, rather than by braking.

Wind Tunnels
Cycling technology is tested in wind tunnels before it is tested on the track. The effects of wind passing moving objects can be tested and then **analyzed** on computers.

The back wheel provides the bike with power. It is a solid disk, because it does not need the blades that the front wheel needs for steering purposes.

The rear wheel and the pedals are connected through a single gear anchored to the rear wheel.

Track bikes have narrow handlebars positioned in the center of the bike. This is to reduce the **drag** that occurs when air hits the front of the bike.

The front wheel is used to steer the bike around the track. It has flattened blades, rather than spikes. This is to reduce drag.

◀ A track bike is made from a variety of materials, but the main material is carbon fiber. Carbon fiber is a very tough, flexible, and light material that can be bent into almost any shape without breaking.

Word Watch

analyzed examined in detail

drag resisting force of air or liquid

24

Road Bikes

Road bikes race on roads, not tracks. Like track bikes, they are made from light materials. The weight of the bike is particularly important during hill or mountain climbs, when pedaling a heavy bike would be far more difficult than pedaling a lighter one.

▼ Road bikes are made from lightweight materials, particularly carbon fiber.

Reducing Air Resistance

Cyclists wear tight fitting, lightweight clothing to reduce air **resistance**, or drag. Clothing that flaps around catches air and slows the cyclist down. In a cycling race, this could mean the difference between winning and losing. Pointed helmets are designed to reduce resistance, too.

Road bikes use a calliper brake system that is centrally mounted to the frame. A calliper brake system uses spring mechanisms that press the brake pads to the rim of the tire, stopping the tire's movement.

Handlebars are positioned lower than the saddle. This forces the rider into a position that minimizes the amount of drag caused by the rider's body.

Gears are designed and positioned so that they can be changed with the least amount of effort during a race.

The wheels have as few spokes as possible. This is to reduce drag.

The tires are light and narrow. They are inflated to a high level to make them as hard as possible. The harder they are, the more speed they can build up.

Word Watch

resistance ability to withstand the effects of something

Web Watch ▼

www.olympic.org/en/content/Sports/All-Sports/Cycling/Cycling-Road

www.olympic.org/en/content/Sports/All-Sports/Cycling/Cycling-Track

How Do Performance-Enhancing Drugs Work?

Scientific and medical technology has been used to create drugs that improve an athlete's performance. Performance-enhancing drugs are illegal in professional sport, but some sports stars are still caught cheating.

Anabolic Steroids

Athletes who take anabolic steroids build up their muscle density, giving them more strength and power. Steroids act in the same way as the **hormone** testosterone. Testosterone encourages the growth of male physical characteristics, particularly muscles. Anabolic steroids are taken by injection or in pill form.

Anabolic steroids may be used illegally by weightlifters, who need strength, and sprinters, who benefit from more power out of the blocks and over short distances. Steroids also enable athletes to train harder and to recover more easily after a race or a hard training session. There are many side effects to steroid use, too. Steroids can:

- increase blood pressure
- increase cholesterol levels, which can affect the heart's function
- damage the liver and kidney
- cause unprovoked aggression
- cause **infertility**

Female Athletes and Anabolic Steroids

Female athletes who take anabolic steroids can grow facial hair, develop a deep voice, and lose hair from their heads. In the 1970s and 1980s, many female athletes from East Germany and the Soviet Union took steroids to build their muscles up in ways that women cannot do naturally. This gave them a huge advantage over competitors who did not take drugs.

◀ Canadian Ben Johnson won the 100-meter final at the 1988 Olympic Games, but had to return his medal when he tested positive for anabolic steroids. Anabolic steroids give an athlete an unfair advantage and can also damage the athlete's health.

Word Watch

hormone substance produced in the body that activates organs

infertility inability to conceive children

26

Amphetamines

Amphetamines aid athletes who are competing in short races or events where short bursts of energy are required. They stimulate the body's nervous system in the same way that the hormone adrenalin acts when a person's body is preparing for activity. They increase heart rate and breathing rate, make people more aware of their surroundings, and speed up their reflexes. They can also cause damage to the heart, weight loss, and sleeplessness, and make it difficult for people to concentrate.

Human Growth Hormone

Since the 1980s, some athletes have used human growth hormone (HGH) to build up body size and strength. HGH is produced naturally by the body, but chemists have found a way to create it artificially, too. It is most commonly used to treat children who are well below average size and who are not growing naturally.

Blood Doping

Blood doping is a medical procedure that is banned by sporting authorities. It involves removing blood from an athlete several weeks before an event, separating the red blood cells from the rest of the blood, storing these cells, then injecting them back into the athlete shortly before an event. Red blood cells carry more oxygen than other parts of the blood, and the more oxygen an athlete has in his or her bloodstream, the better the athlete's performance over a long distance or period of time.

World Anti-Doping Agency

In 1998, during the Tour de France cycle race, a raid found a large number of cyclists on performance-enhancing drugs. These drugs were seen to be damaging sport. In 1999, the International Olympic Committee set up the World Anti-Doping Agency. This organization leads the fight against performance-enhancing drugs in sport.

▲ A scientist tests athletes' urine samples for banned drugs.

Web Watch

www.wada-ama.org/en/dynamic.ch2?pageCategory.id=312

www.howstuffworks.com/athletic-drug-test.htm

How Do Pole-Vaulters Launch Themselves So High?

Pole-vaulters use a long pole to launch themselves into the air and over a bar that is several feet above the ground. Technology has been used to develop new, lighter poles and also to help coaches and athletes **analyze** performances and improve technique.

Vaulting Across Distances

Pole vaulting occurred for hundreds of years in Europe before becoming an official sport. It began as a way for people to get across narrow bodies of water or boggy ground. It became an official sport in the 1800s. At first, the idea was to vault the farthest, rather than the highest.

Word Watch

analyze examine something in detail

Web Watch

entertainment.
howstuffworks.com/
pole-vault.htm

The Vaulting Pole

When the pole vault was first invented, poles were made from bamboo. Bamboo provided athletes with the bend and spring they required to reach the bar. As techniques improved and pole-vaulters jumped higher, these bamboo poles were replaced with stronger, steel poles. Today, vaulting poles are made from layers of fiberglass, including lightweight carbon fiber. These poles are as strong as steel poles but have greater flexibility.

A carbon-fiber pole is light to carry on the approach, when the pole-vaulter is running with the pole. This maximizes the speed that the vaulter can reach before launching into the air.

◀ A vaulting pole must absorb a pole-vaulter's energy while it is bending, then release that energy back to the vaulter as the pole straightens out. The heavier the vaulter, the more stress the pole is under. So, a heavy vaulter needs a stiffer pole than a lighter vaulter.

How to Pole Vault

There are six main stages to a pole vault jump.

 Approach

The pole-vaulter starts her run-up holding the pole in a **vertical** position. During the run, she lowers the pole so that it becomes **parallel** to the ground and chest high.

 **Plant**

The plant is the moment that the bottom of the pole is planted in the box at the base of the jump.

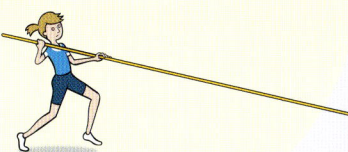

 **Takeoff**

The pole starts bending. At the moment of takeoff, the vaulter jumps off the back leg and drives her front knee forward, **propelling** herself into the air.

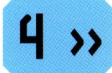 **Flight**

To gain maximum height, the vaulter maintains a low **center of gravity** while the pole continues to bend, then transfers her center of gravity forward when the pole jumps back.

 **Turn and Clearance**

The turn is put off until the pole-vaulter is as close as possible to being completely vertical to the bar. At this point, there is no more spring to aid the vaulter. The vaulter faces the bar and pushes the pole away from the bar.

 **Fall**

Once she has cleared the bar, the pole-vaulter pays particular attention to propelling her arms away from the bar to make sure she does not touch or bump the bar.

Greatest Pole-Vaulter

The greatest pole-vaulter of all time is Sergey Bubka (1963–) from the Ukraine. He competed in major championships from 1981 to 2001. Bubka broke 35 world records, and he was the first pole-vaulter to clear a height of 19.69 feet (6 m).

Word Watch

center of gravity point where an object's or person's weight will evenly balance

parallel side by side, with the same distance between them

propelling pushing forward

vertical at right angles to the horizon

How Do High-Tech Swimsuits Give Competitors an Unfair Advantage?

In swimming, the difference between winning and losing can come down to $1/100$ of a second. **High-tech** swimsuits that reduce **drag** in the water have been developed, but this technology is now banned because it gives too much advantage.

Conserving Energy
By wearing a swimsuit that reduces drag in the water, a swimmer has more energy toward the end of a race when the swimmer needs it most.

Understanding Drag

In swimming races, drag is the resisting **force** that water exerts on a swimmer as the swimmer powers through the pool.

▶ Three types of drag occur when a swimmer powers through the water.

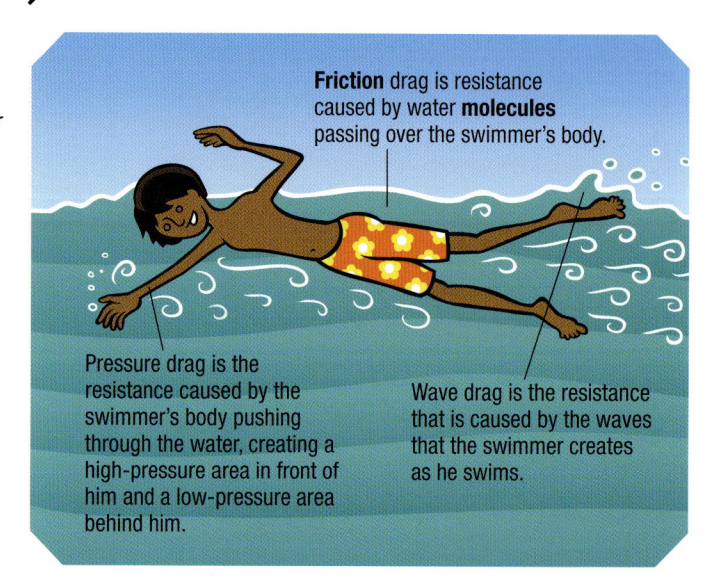

Friction drag is resistance caused by water **molecules** passing over the swimmer's body.

Pressure drag is the resistance caused by the swimmer's body pushing through the water, creating a high-pressure area in front of him and a low-pressure area behind him.

Wave drag is the resistance that is caused by the waves that the swimmer creates as he swims.

Development of the LZR Racer®

The swimsuit that caused the most controversy is the Speedo LZR Racer®. It was developed by Speedo with the help of many high-tech organizations, including the National Aeronautics and Space Administration (NASA), and it underwent years of testing during its development. Testing included:

- surface drag testing—Using NASA's wind tunnel technology, Speedo tested the surface drag of more than 60 types of fabric.
- water flume testing—This took place at the University of Otago, New Zealand, in a narrow, 33-foot (10-m) long channel of precisely controlled flowing water. Both **mannequins** and real swimmers were used to test various suits.
- body scanning—More than 400 elite athletes had **three-dimensional** scans of their bodies produced so designers could learn more about the shape of swimmers' bodies.
- computational fluid dynamics—This involved using computer modeling technology to predict how designs would behave in real-world environments.

Word Watch

drag resisting force of air or liquid
force a push or a pull
friction force that resists motion and slows an object
high-tech (high technology) advanced
mannequins dummy of a human
molecules group of atoms stuck together
three-dimensional using length, width, and depth

High-tech Features

The fabrics used to make high-tech swimsuits are extremely light and strong. They are also water repellent and fit tightly around the swimmer's body. As a result, drag is reduced.

There are no stitches, because stitches cause drag. The various pieces are **bonded** together using high-tech fabric-welding techniques.

A strong central support holds the body in.

The suit's zipper is bonded into the suit so it maintains its smooth, sleek surface.

Ultra-thin, low-drag panels are positioned at points on the swimmer's body to reduce drag.

▲ High-tech swimsuits fit snugly to swimmers' bodies.

Ban of High-Tech Swimsuits

In 2009, the world governing body for swimming, FINA, introduced new rules that banned high-tech swimsuits such as the Speedo LZR Racer®. Many records were being broken by swimmers in these new suits. FINA was concerned that technology was taking over from natural ability in determining who won races.

> *FINA wishes to recall that the main and core principle is that swimming is a sport essentially based on the physical performance of the athlete.*

FINA, the world governing body for swimming, when it announced its ban on high-tech swimsuits in 2009

Word Watch

bonded stuck together

Web Watch

www.speedo80.com/lzr-racer

Index

A
accidents 16, 17
Adidas running shoes 6
aerodynamic forces 21
air resistance 25
amphetamines 27
anabolic steroids 26
auto racing 5, 16–17, 18–19

B
bicycles 24–5
black boxes 17
blood doping 27
Bolt, Usain 8
Bubka, Sergey 29
Burke, Thomas 8

C
computer simulations 9, 17
cricket 12
Cyclops system 13

D
Dassler, Adi 6
drag 20, 24, 25, 28, 30, 31
drag racing 18

F
FINA 31
fire extinguishers 16

G
Gates, Bill 4
golf balls 5, 20–21
golf clubs 21

H
harnesses 16
Hawk-Eye system 5, 12–13
Hawkins, Paul 12
heart 10
heart-rate monitors 10–11
human growth hormone (HGH) 27

I
International Olympic Committee 27

L
lawn bowling 14–15

M
marathon running 8

N
National Aeronautics and Space Administration (NASA) 31
Nike running shoes 6
nutrition 5, 9

O
Olympic Games 8, 9, 23, 26

P
performance-enhancing drugs 5, 26–7
plimsolls 6
Polar Electro 10
pole vault 28–9
protective clothing 17

R
racetracks 5, 16, 17
radar gun 13
Reebok running shoes 6
road bikes 24, 25
running 6, 7, 8, 9
running shoes 6–7

S
seat belts 16
sensors 6, 7, 16
slicks 18
snowboards 5, 22, 23
soccer boots 7
Speedo LZR Racer® 30, 31
swimsuits 5, 30–1

T
tennis 12, 13
Tour de France 27
track bikes 24
treads 18, 19
tires 18–19

V
vaulting poles 28, 29
video technology 9

W
wind tunnels 24, 31
World Anti-Doping Agency 27

MAY 1 3 2011

J 688.76 B736
Brasch, Nicolas.
Sports and sporting equipment /